BIG TREES & STEAM LOKIES
AND A LOT OF BULL TEAMS

Written By

DAVE JAMES

Dedicated to Albert Wiesendanger, who gave most of hs 98 years to progressive forestry, Keeping Oregon Green and hunting for BIG TREES.

YE GALLEON PRESS
FAIRFIELD, WASHINGTON
1991

First Printing, May, 1989
Second Printing, June, 1989
Third Printing, April, 1991

WAS IT THE BIGGEST EVER?

THE NEBY RIVER FIR

This booklet has one purpose: to recall through previously unpublished photographs some of the glory days of logging in Western Washington when one log made a load. Some of the biggest Northwest trees ever felled went to the mills in this century after Washington became a state.

Instead of one huge stem on truck or rail car bunkers, we now see 15 or 20 or more logs of the new forest called Second Growth. This is the future way tree needs of the nation will be served by fast growing cycles of man-planted seedlings.

Robert Sheane was a photographer at Shelton, Washington, in 1903. He is gone and likely forgotten except for his pictures of a tree which may have been the largest Douglas Fir ever to grow in Washington.

Sheane and his glass plate camera box arrived on the Neby River west of Simpson Logging Company woods headquarters at Matlock none too soon. Fallers were already chopping an undercut in the mammoth trunk on this summer day in 1903. Someone who had realized the unusual size of the tree must have called Sheane to come and "shoot it."

Sheane took the conventional scenes. Fallers on their spring boards with axes and cross-cut saw. Fellow loggers standing against the craggy butt. Loggers posed in the undercut, daring the tree to bite them. And, of course, the great butt log of the fallen monarch.

After the Neby Fir crashed to earth, a log scaler measured its diameter *inside the bark* at 14 feet 11 inches. This measurement was painted on the pitch-bleeding butt and remained on view several years. That butt log never moved from the stump. It was too big to pass through a railway cut to reach the Shelton dumping grounds.

"Inside the bark" is the key to the question: Was this the largest Douglas Fir ever to sprout from a seed so tiny it takes 40,000 to weigh a pound? Official measurements of champion trees are taken outside the bark. Sheane's pictures reveal the bark was thick, easily six inches and likely more. This would have increased the diameter by a full 12 inches to 15 feet 11 inches.

Tree judges measure the trunks at diameter breast high, 4½ feet above the ground. The Neby Fir was measured several feet higher where the trunk was severed. This was two boards high, well above DBH, and above the normal swell which would have added more inches.

Did the Neby Fir actually measure between 16 and 17 feet? We'll never know for sure. Its stump has been leveled to a mound of compost by nearly 90 years of nibbling, sucking and chewing by ants, bugs, grubs and decay. This intriguing question inspired BIG TREES and STEAM LOKIES as a reminder of Washington woods giants never to be felled again.

--Dave James

In the redwood country of Northern California twelve logs formed a train of twelve cars, all that the Baldwin lokies of 1890 could pull to the mill. Simpson Timber Company is growing young redwood forests on lands near Arcata where these logs were stripped of their bark and shipped out like great loaves of bread.

–Simpson Archives

This tree standing on the Neby branch of the Satsop River west of Shelton, Washington, may have been the largest Douglas Fir ever felled. Photographer Robert Sheane posed the fallers as they were chopping the undercut in the summer of 1903. The spectators were fellow loggers. When cut down, the butt log was measured at 14 feet 11 inches "inside the bark," which would have added at least another foot to the diameter, possibly more.

–Simpson Archives

Word spread quickly through Simpson camps that a helluva big tree had been dropped at Camp Five. Families decked out on the Sunday following to inspect the tree and gasp at its size. Buckers were already cutting it into lengths that could be managed by the equipment available in 1903. The butt log had to be left where it fell. It was too wide to pass through a narrow cut on Simpson's railroad. Bugs eventually ate up the butt.

—Simpson Archives

This was the cut, 90 feet deep, through which Simpson trains were able to haul all of the Neby River tree except its thick butt log. Widening the cut for one log was too costly.

—Simpson Archives

"Stump standing" became a form of competition between Simpson camps after the great Neby River tree hit the ground. One Sabbath, 24 well-togged-out loggers and a child congregated on the stump, leaving six people below to await their turn to climb the ladder. The ladder looks flimsy but we can assume all stump standers descended safely.

Simpson Archives

Not to be outdone, a posse of tough timberbeasts posed with a horse and an armed leader on this same stump, marked Camp 5. It is easy to imagine the snorts heard in the bunkhouse when it was said the horse jumped upon the stump. Well, how did it get up there?

--Simpson Archives

When the first log moved from the Neby Fir reached Shelton, it was dumped off downtown to be bucked into firewood. Upper Puget Sound mills buying logs from Simpson said it was too big for their sawmill carriages. Robert Sheane lugged his camera over to record this next phase in the Big Tree story. Old Whiskers over beyond the first cut off the log may have been about to ask if he could buy a rick of stovewood for his kitchen range.

–Simpson Archives

Editor Grant C. Angle of the weekly Shelton Journal knew how to attract subscribers to the paper he started in 1886. With a slice of the Neby Fir on display, Angle hired this little man with the big watch chain to sit up there and wave a sprig of apples at passers by.

–Simpson Archives

Most of the woods photography performed by the brothers Darius and Clark Kinsey dealt with chopped down trees, working animals, machines and loggers. On this occasion Darius sensed the beauty of the uncut forest and posed a woodsman peering wistfully into the distance. These trees are typical of the timberlands which thrived where Puget Sound cities are now blotting out logged off countryside.

—Simpson Archives

Giant Douglas firs of this size sometimes grew so close together a fat-bellied man could not squeeze between them. Loggers working on such trees respected their size and sometimes counted their age rings to determine the centuries they had lived. The man with outstretched hands wants you to know he's never caught a fish this big.

–Simpson Archives

Clark Kinsey, who recorded so much logging history in his thousands of negatives now preserved at the University of Washington, took time for a picture of himself and son, Ronald, with this craggy-barked 11-foot Douglas fir near Simpson Camp Five in the 1920's.

—Simpson Archives

This portrait of a super-sized Douglas fir hung in Simpson Logging Company offices in Shelton for a half-century. Unknown are the soft hatted fallers or the city man taking notes on who took the picture. Note the lower springboard holes which enabled the fallers to work "two boards high" where the tree bind was not so resistant to the saw. Estimated diameter: 12 feet.
–Simpson Archives

Darius Kinsey failed to record where in Western Washington he pictured this giant Douglas fir in 1905. He did note the tree measured 52 feet in circumference at four feet above the ground. There is noticeable swell in the trunk to that height but this obviously was one of the biggest firs ever felled. The fallers are in classic pose with axes and saw while a fellow worker reclines in the undercut, relaxing while the boss is not around.

—Simpson Archives

Taking a seat in the undercut of this girthy Douglas Fir would not be pleasing to modern Yuppies decked out in Brooks Brothers garments. How these three got away from the tree without being pried or doused with saw oil is a good question. Rivers of sticky pitch usually flowed when a big tree was chopped. Some fallers carried pitchy wood chips home to start a kitchen fire. Chips, oozing pitch made wives so happy, loggers called them "woosticks."

–Simpson Archives

Falling a six-foot tree meant several hours of chopping and pulling the crosscut saw. The members of this 1940 Simpson cutting crew rightfully show pride in the glass-smooth undercut they have chopped. Modern power saws have replaced the axe-blade skills which once were the mark of an experienced faller. The best could chop right or left handed.

—Simpson Archives

It may have been a day's work for the man sitting on this huge log to get up on it. The average sized man standing below shows this tree, felled near Matlock, Washington, around 1900, was nearly 10 feet in diameter high up the trunk. Surely a magnificent tree. The biggest Douglas Firs grew on bottom lands near streams.
--Simpson Archives

As he prepares to go to work, our log bucker seems to be contemplating how many saw strokes it will take to cleave this imposing Douglas fir log. Or it could be he has just heard a black bear woofing on the other side.

Simpson officials summoned Photographer F. Billie Wilcox all the way from Centralia, 45 miles away, to get a shot of Scalers Dave Adams and Al Klasell running a tape across this massive log hauled from Camp Five in the Thirties. When logs were too big for sawmills it was necessary to pop them apart with a touch of dynamite.

--Simpson Archives

Relaxing in this freakish, never-to-be-duplicated four-pronged Douglas fir is Will Grisdale, the Simpson logging boss whose crews cut four billion feet of timber during his half century with the company. This one of a kind tree was cut down early in the century. Each "sprout" was in itself a big tree. Deer browsing when the tree was a seedling likely deformed it.

–Simpson Archives

Lean, muscular, flick-quick and no man to meddle with. Call him bullwhacker or bullpuncher; he didn't mind. Just so you respected his abilities and authority. Not a load of shingle bolts or a string of logs budged until he gave the command.

Standing by his "wheelers," always well-endowed bulls, this "puncher" of the 1880s demonstrates affection for his animals. True, he could roar and jump and shout like a sideline coach, but he really cared for his plodding beasts. He saw that they were well fed, housed in dry barns and given care when their labors left them lame.

Most of the animals used in logging were oxen — castrated adult bulls — but the wheel teams which started the pulling effort were bulls trained to respond to stop and go commands. The common name was "bull team" even though oxen were always present. —Simpson Archives

Most bulls and oxen were red, black or splotched white and gray. Another term for mixed coloring was "brindled." This elegantly-horned seven span team included a rare white oxen. The bullwhacker stands behind Whitey, holding his slender goad at ease. Behind him worked the hooktender, who supervised hitching the logs with chains and "dogs." The hand-skidder was the hooktender's helper. He carried a big maul to drive the dogs into the logs. To remove them he used the handle as a pry. Others were fallers and buckers who had put down their saws to get into the picture. Bull teams moved quietly, chains clanking, and they moved at only one speed no matter how loud the bullwhacker yelled to the Almighty for more action.

—Special Collections, U. of Washington Libraries

This rare photo of a bull team pulling logs loaded on wagon wheels likewise shows the girthy Douglas fir forests in which Washington loggers worked in the 19th century. Logs weighing many tons were supported by four wheels up front and four arrears (not shown). Four yoked teams (also called spans) could pull the load on a level grade. The lead bulls are open-mouthed. Possibly they were bellowing for hay, or they may have been burping. Several animals are polled (no horns). Usually the bull teams wore long flaring horns, capped with brass knobs for safety reasons. Getting goosed by a horn was the worst dream a bullwhacker could have.

—Special Collections, U. of Washington Libraries

Bulls shod with iron plates nailed to each cloven hoof were kept in corrals and barns at night. Overhauling of footwear was done on Sundays or days when the weather was too rude for logging.

The animals were yoked in pairs. A U-shaped collar, usually made of maple, was fitted around the neck and thrust through holes in the heavy wood yoke. Pins held the collar in place. Yoke carving was a skill calling for smooth shaping of the wood to avoid galling the animal's neck. Loggers who were concerned about the health of their animals would not work them in wet weather. Wet yokes caused neck sores.

Billy Forbes was a famous Mason County bullwhacker. He started when he was 19 in 1883. Author Stewart Holbrook pried these words out of Billy when he wrote GREEN COMMONWEALTH in 1945.

"A good teamster worked a 16-hour day for $100 a month and board. The bulls weighed up to 1800 pounds. Dehorning wasn't done usually but sometimes while working a horn got knocked off.

"Accidents to bulls were not common. In a dozen years I never injured but one bull seriously. A log made a quick run on a turn and broke the bull's leg. (This was costly because bulls sold for $500 to $1,000. The cost was partially offset by the steaks and hamburger salvaged by the cook.)

"Bulls worried about nothing but yellowjackets. Hornets sure would raise hell for both the bulls and the whackers. No cruel bullwhacker lasted long. I commanded respect with a goad stick five feet long. It had a one-inch brad in one end. I poked my bulls only to get their attention."

—Mason County Historical Society

This camp scene was typical of the bull team years. The bulls are resting after having dragged eight heavy logs to a rollway where flatcars (rattlers) were loaded. The front end of each log has been sniped (tapered) to avert hangups as they slid over greased logs forming the skidroads. The scene includes the all-important blacksmith shop (foreground) where hooks, chains and bullshoes were shaped. Up the hill is a saw filer's shanty and in the background a tool shed. The steam donkey, lower right, pulled rail cars into loading position. Stumps eight to 10 feet high were a common sight 100 years ago. Fallers stood on springboards "two boards high." The flow of pitch in Douglas fir giants was less of a problem to saw pullers at that height.

—Duane Dewell Collection

Five or six spans of bulls and oxen usually supplied all the pulling necessary to moving logs a mile or more. This team of 11 yoke (22 animals) may have been required to drag logs on an uphill pull. The crew of six men standing on the lead log were not riding into camp. They were posing for the photographer. Careful readers will note there are two bullwhackers, one by the wheel team hitched to the log and another midway up the procession, waiting to whisper into the ears of his horned friends.

—Duane Dewell Collection

How many years did a bull or ox last at its hard life in the woods? Quite a while. Levi Shelton owned 11 yoke of oxen in Mason County in 1888. They ranged from five to 10 years of age and had names like Tom, Bright, Lion, Jerry, Brandy, Bully, Nig and Turk. Bullwhackers spoke to each by name when urging greater effort. Old loggers snickered when they said "Get in there and pull or you'll be in tomorrow night's stew." Of course they didn't mean it because a healthy animal was worthy anywhere from $500 to $1000.

—Mason County Historical Society

Bulls must have sensed their time was nearing the end when they first saw this primitive steam donkey pulling a string of railroad cars. This location was called a rollway. It was elevated just above the flat cars or "rattlers." As bull teams pulled logs into position, burly men wielding peaveys inched the logs onto the railcars. This method was used before loggers developed equipment to lift and lower logs. Bull team logging lasted into the mid-1890s.

—Duane Dewell Collection

Not all railroad tracks meant a locomotive was puffing away at hauling logs somewhere. To bring its big logs down a 4000-foot incline on a 48 percent grade, the McCleary Timber Company eased loads down with a heavy line unreeled by a donkey engine at the top of the grade. A box at the right of the rear car contained a telephone. The donkey engineer would phone to say "I'm letting her down." The chaser at the bottom would yell into the mouthpiece so loud he did not need a phone "Easy, George, the last one damn near went to town."

–Simpson Archives

Sol G. Simpson sent this "Shelton toothpick" to the Chicago World's Fair in 1892. Containing 14,000 board feet, it still holds a record as the biggest piece of a tree ever hauled to the Middle West from the West Coast.
—Simpson Archives

Single log loads were a common sight at Simpson's Camp Five in the 1930's and 1940's when some of the largest Douglas Fir trees ever cut in Grays Harbor County were rail-hauled into the booming grounds at Shelton. Besides industrial and home building uses, a strong market for straight-grained Douglas fir came from Italy, where the centuries-old wood was sliced into thin and narrow slats for Venetian blinds.

–Simpson Archives

Stock-piled logs, called cold decks, awaited loading on flat cars. Modern safety supervisors would turn as pale as a smelt's belly at the sight of the unprotected "cat" driver and workmen standing among the tangled logs. This was a common woods practice 50 years ago. State safety regulations finally stopped "Cold Decking."

—Simpson Archives

Douglas firs and Sitka spruce vied for record size in the Pacific Northwest. Both were outmatched by California's redwoods, the larger but fewer Sequoia Gigantea in the mountains and the taller Sequoia Sempervirens along the Northern California coast. Clark Kinsey came upon this 22-foot undercut in a Coast redwood near Eureka just as two of the cutting crew had stretched out for a nap. Trees of this size are gone from commercial forests but big ones can still be seen in California's numerous redwood parks.

—Clark Kinsey

This exceptional photo of bull teams at work in a Washington logging camp on the Columbia River in 1897 is one time the Kinsey brothers got scooped. A little-known Seattle photographer named Eakins perched his cumbersome lens box on a hillside for this view. Three bull teams are bringing logs to a rollway for car loading. Men in camp have swarmed out of the cookhouse to get in the picture, leaving the cook to rant and rave.

–Simpson Archives

The baleful eye of the bull chained in this sturdy laboratory of the bullshodder lets us know it was humiliating to have one's tail held like a rudder while the smithy nailed on iron shoes. One can assume the angry bull was pondering whether to let them both have it full blast. The little man holding the goad is half grinning at what he hoped might happen. If it did, he wasn't out of range himself.

—Special Collections Division
University of Washington Libraries

If inventor James Watt had ever seen the Herculean strength of bulls pulling enormous logs over skid roads in the 1890's, we surely would speak now of engines having bullpower instead of horsepower. These 12 animals yoked in pairs were moving a Douglas fir log 12 feet in diameter and weighing some 50 tons. Starting the pull meant a rippling movement from the wheelers to the leaders in a surge of power orchestrated by a bullwhacker using some of the warmest profanity ever voiced by man. Older oxen got accustomed to cursing after a few years.

--Special Collections Division
University of Washington Libraries

Lacking a land route, Sol G. Simpson brought his lokies via barge to Kamilche in the early 1890's. Known as the "Kneeland Lokie," this 2-6-0 type built by Clark in 1881 came to the Shelton Southwestern line in 1895 and was traded to the Blakely line for its "Skookum" in 1898. Hay for Simpson's early animal teams came to Kamilche.
—Simpson Archives

At times the locomotives in Simpson camps did the work of oxen and horses. This 2-truck Heisler type, built in 1896, dragged logs directly over the ties to a rollway where cars were loaded. The U.S. Department of Agriculture's Bureau of Forestry, forerunner of the U.S. Forest Service, recorded this scene at Matlock on August 7, 1899. Logs dropped on the rollway were shunted onto waiting "Rattlers" by pure sweat and fury.

--Simpson Archives

This is a sight that, seen by a soused timberbeast just back in camp from an Aberdeen weekend, would have put him on his knees vowing never to touch the wicked juice again. Track inspectors followed logging trains in dry weather to make sure sparks had not set the woods afire. Pedalcars were light enough to toss off the tracks if an unscheduled train of empties came screeching around the bend, which sometimes happened.

—Washington State Historical Society

Another example of a log-dragging lokie, this 35-ton Climax built for Simpson in 1898 is draped with hooks and cables it used for pulling logs over the ties. The two-truck geared wood-burning engine was named "Irene" for Sol Simpson's daughter, who married Mark E. Reed in 1901. Reed later became one of Washington State's leading industrialists and politicians. In their heyday Simpson Camps had 13 lokies and 300 miles of track.

–Simpson Archives

Contrast the bulls at work in 1897 with the fleet of "Cats" Simpson had dragging logs at Camp Five in the Thirties. Mechanization enabled loggers to hustle out logs from elevations impossible to reach when plodding bulls and lunging horses strained at their exhausting burdens. After some accidents the drivers were shielded.
–Simpson Archives

Now and then the rails spread apart, tipping a locomotive to expose everything it had underneath. The engineer and fireman leaped for their lives when this proud creature upset into a ditch. Examining the derailment were a dozen loggers out from Simpson's camp near Matlock. It was Sunday and the men dressed for the occasion. Despite serious upsets few railroad workers were hurt in wrecks. They knew when to leap into the bushes.
--Simpson Archives

Steam power moved the earth and built the rights of way long before combustion engines were introduced into Washington woods. Only a Kinsey would have thought of catching this scene of a steam shovel working while a gravel train waits to be loaded. In the background stands a water tank from which the steamers drank deeply and often. Water tanks were spaced at regular intervals along the tracks. The flower pot stack controlled sparks.

—Clark Kinsey

Wives with a taste for interior decoration influenced at least some of the grizzled and growly old-time Simpson engineers. What else could account for the starched window curtains in the cab of this 2-6-2 rod engine blowing smoke beside the company shops on Railroad avenue in Shelton in 1905? Loaded with split wood for its next run to camp, road engines of this type also carried sand to spill on slippery steel when rain-soaked Autumn leaves "greased" the upgrade tracks. Contractors piled firewood beside the tracks every few miles.

--Simpson Archives

Young Ed Elliott, seated at left on the frame of the Six-Spot, later became Simpson's first railroad superintendent. His fellow workers are unidentified. Baldwin built this 2-6-2 "Prairie Type" lokie in 1906.

After Simpson acquired larger and more powerful engines, the Six-Spot was assigned to construction and other lighter work. Shay-type Lokies became more frequent as loggers moved into the mountains.

--Simpson Archives

Mark Draham, a famous boss logger in Mason and Thurston counties early in the century, bought this boxy saddle tank lokie for his Western Washington Logging Company in 1908. The 2-6-2- type was built by Baldwin as a wood burner but was later converted to oil. Its later years were with the Mud Bay Logging Company near Olympia. An advantage offered by the saddle tank was more weight over the drive wheels.

--Simpson Archives

The Eight-Spot, a 2-8-2 Mikado type engine built by the Baldwin Locomotive Works in 1910, hauled logs from Simpson camps for 46 years before going to a scrap heap in 1956. Originally a wood burner, the 86-ton engine was converted to oil in 1916. The little building atop the tender had nothing to do with sanitary convenience. It was the "dog house" in which the front-end brakeman stood out of the chill and rain as he kept an eye on loaded flatcars snaking behind the lokie. Posing in 1950 were, from left, Ed "Junior" Whinery, brakeman; Vic Ellison, crew member; Clay Berry, railroad superintendent; Arnie Aronson, engineer, and Lawrence Larson, conductor. It was not unusual for men to work 40 years or more on Simpson's line.

–Simpson Archives

Train buffs got bright eyes looking at the 2-8-2-Mikado, the Three-Spot, pulling logs from Camp Grisdale into Shelton. Named after William Shorter, the engineer who ran Simpson's original Three-Spot on the Satsop and Peninsular lines at the turn of the century, this was the last steamer to be built for Simpson. It was scrapped in 1957, after Simpson replaced its steam fleet with four diesel electric lokies.

–Simpson Archives